# ENVIROlocity™ Job Search Workbook

## A step-by-step guide to creating an action plan that will motivate you and accelerate your ideal environmental career search

## ENVIROlocity™

*Master the three keys, accelerate your job search, and stay motivated!*

# Contents

*Fillable worksheets and templates available online at
theenvironmentalcareercoach.com/workbook*

# Welcome!

Congratulations! Just by reading the information provided here, you have chosen to take charge of your life and your career rather than take a passive back seat. You are contemplating what it might mean to take your job search seriously. Creating a plan will make your job search less stressful, more focused, and even more enjoyable.

Overall, as businesses incorporate environmental awareness and responsibility into their strategies, environmental job opportunities are increasing. Despite that fact, some environmental graduates never even start a career in an environmental field. Others start their environmental careers but don't stay in them long. Most often, this happens because of unmet expectations or too many workplace pressures. My goal is to ensure that students and career changers understand the challenges they face, know what to expect, and are equipped for the road ahead.

## The purpose of this workbook

"A fool with a plan is better off than a genius without a plan."
-T. Boone Pickens

While there is no guarantee that by reading this workbook you will find a job in x number of days, I can guarantee that if you put real thought into each recommended activity in this handbook, you will end up with a solid plan to help you make better, faster, and more confident steps toward the right career for you. You will not be disappointed with your results.

Note that throughout this workbook, when we refer to jobs or job search, we are referring to the job that will be the next step in your environmental career. We encourage you to think of your career goals and not "just a job".

While I'm inclined through habit to wish you good luck, I prefer to advise you to get out there and make your own luck!

-Laura, The Environmental Career Coach

# How To Use This Workbook

The purpose of this workbook is to walk you through the necessary steps to create a robust and effective action plan for your job search. The exercises will help you get clarity about the type of career you want, answer your questions about what to search for, and complete an action plan that you will use to stay accountable to your career dreams.

The workbook follows the Environmental Career Coach's 3 Keys to an Effective Job Search model. The exercises are designed to get you to write down ideas, identify gaps, and conduct research. The results from the exercises will become the basis of your action plan. Each section contains two exercises. Read through the intro of each exercise before getting started. Each exercise is intended to help you generate ideas for potential action items to add to your job search action plan. You'll store your ideas on the holding page (page 33) as you go. Be sure to set time aside to do each exercise and not rush through them. Allocate 30 minutes for each exercise and add more time if needed.

From our experience, an effective action plan includes goals in three areas: networking, applying, and learning. Balancing tactics in these areas ensures that you don't put too much attention on just one aspect of career planning, to the detriment of the others. After compiling your ideas from each exercise, you'll sort and prioritize them into the three buckets on the sorting page (page 34). There are extra notes pages in the back of the workbook; if there is not enough space use another sheet of paper to write your responses. You can also replicate the worksheets in a spreadsheet or access spreadsheet templates and copies of the worksheets at theenvironmentalcareercoach.com/workbook.

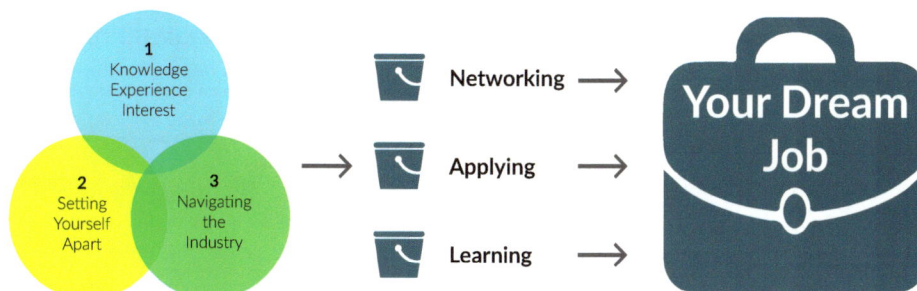

Lastly, make sure that you invite someone to be your accountability partner and give them access to your action plan so they can give you feedback and check on your progress. This person can be a reliable friend or sibling, a fellow student, an advisor, or a coach.

Take your time going through this content. Consider setting some time aside every day for a few weeks to put real thought and effort into the exercises.

# Start With The ENVIROlocity™ Mindset

If there's one thing I've learned in the time I've been career coaching, you are your own worst critic and deplorable cheerleader. Learn to recognize when you are getting in your own way.

|  |  |  |
|:---:|:---:|:---:|
| *Self-Defeating Mindset 1:* | *Self-Defeating Mindset 2:* | *Self-Defeating Mindset 3:* |
| **Preconceptions** | **Self-pity** | **Self-Doubt** |

**Preconceptions**
- I'm too old
- It's too hard
- It's too late
- People will think I'm crazy

**Self-pity**
- There must be something wrong with me
- I'm not qualified

**Self-Doubt**
- What if I pick the wrong thing?
- What if I can't get a job?

Unfortunately, it isn't realistic to try to make these feelings go away all together but there are several things you can do to minimize them.

1. **Journal.** Write about your feelings during the job search paying close attention to your inner dialogue and negative thought triggers. Check out our journal "Environmental Job Search Companion Journal: Staying Sane While Seeking Work Worth Living." Find it and more at theenvironmentalcareercoach.com.

2. **Create goals and track your progress towards them.** Actively managing your search will help you to feel more in control. You're already on your way by going through the exercises in this workbook.

3. **Get an accountabilibuddy** who can help you maintain focus on your goals, evaluate your progress, and talk through your self-doubt. A good accountability partner can be a friend, a fellow student, a trusted colleague, a sibling, a mentor, or a coach.

# 3 Keys To An Effective Job Search Model

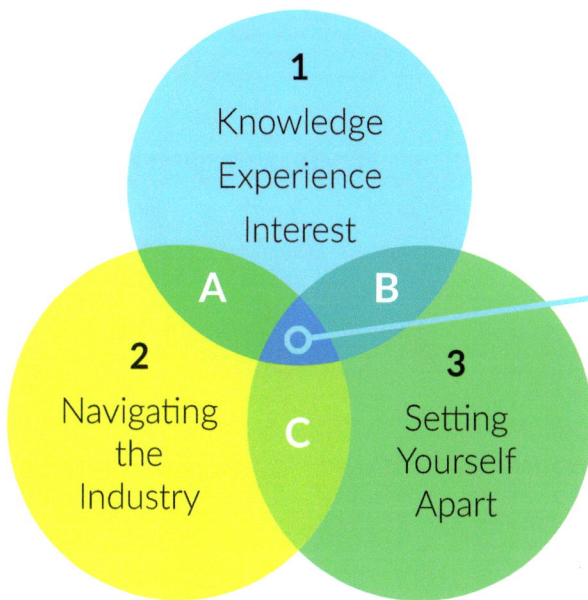

**1**
Knowledge
Experience
Interest

**A**    **B**

**O**

**2**
Navigating
the
Industry

**C**

**3**
Setting
Yourself
Apart

**ENVIROlocity™**

*Master the three keys, accelerate your job search, and stay motivated!*

### Scenario A:
### Path to frustration

If you focus on Knowledge, Experience, and Interest (Key 1) and navigating the industry (Key 2) but neglect Key 3, you may find that you have the passion and know where to apply for jobs but you won't stand out and therefore will not get hired.

### Scenario B
### Driven but no direction

If you focus on Knowledge, Experience, and Interest (Key 1) and setting yourself apart (Key 3) but neglect Key 2, you may have the passion for the jobs you are applying for but they may not be the right jobs. You may be hired for a job only to end up unsatisfied with that career.

### Scenario C
### Stuck in second place

If you focus on navigating the industry (Key 2) and setting yourself apart (Key 3) but neglect Key 1, you will be applying for the right jobs and making a good impression but you will likely not be hired for lack of experience, or you may be hired but lack passion for the job.

# Key 1: Knowledge, Experience & Interest

Align your knowledge, skills, and experience
with the jobs that fit you best.

"Passion is what gets you through the hardest times that might otherwise make strong men weak, or make you give up."
- Neil deGrasse Tyson

The first key to getting started in a career that you will love is finding one that is a good fit for you. The best fit is one that intersects with your knowledge, experience, and interest. This is also referred to as your passion. Don't be afraid to seek your passion. On the flip side, it is okay not to know what your passion is just yet. It often takes several years of trying new things to find out what truly lights you up.

The exercises in this section are meant to help you get clarity around the type of work you'd consider what we call "worth living". Work worth living is the type of job that feels like it was made for you instead of making you dread Mondays.

The first step is taking the time to figure out what you want your life and work to look like in the future. Finding a job that fits your vision is a more sustainable approach to your career than trying to fit your life to the job.

Once you have a clearer picture of your future to use as a guidepost, you'll need to get a clearer picture of where you are now. In the second exercise in this section, you'll review job descriptions for positions you'd potentially be interested in and look for knowledge, skills, and experience that you are missing.

Note that few people can check the box on each and every requirement for the job that they want. The exercise is intended to help you determine which tactics you might add to the learning section of your Job Search Action Plan.

**Exercises included in this section: Visioning & Skills Gap Analysis**

# Visioning Exercise

*"Your vision will become clear only when you can look into your own heart. Who looks outside, dreams; who looks inside, awakes."*
*- Carl Jung*

Have you ever really thought about what your future will look like? You probably have, but you might not have spent time thinking about how much your career influences that future.

Even if you know the exact job you want or have an idea, it is important to consider what kind of life you want to have. Many factors affect your level of satisfaction and happiness you will have including career, industry, sector, location, relationships, income, etc. Do you want to travel? Be the boss? Own a company? To have a successful career, you'll want to consider these things.

The purpose of this visioning exercise is for you to really think hard about your life goals and what your future will be like. Visualize the lifestyle you'd like to have, how much money you'd like to make, if you will have a family, etc. Of course, there are no guarantees about the future but you'll have a better chance of living the life you want if you take control rather than let life take you where it will.

The pages are divided so you can think both about your working life and your personal life at the same time. In order to find Work Worth Living, you'll want to consider both.

## ·········· Instructions for the exercise ··········

1. There are three parts of this exercise: your current situation, a few years ahead, and many years ahead. You can complete them one at a time or all in one session.

2. Find a quiet place and plan to spend at least 10 minutes to answer each section as fully and honestly as possible. Questions have been provided to help guide your thoughts on the example page. You do not have to stick to them and some may not apply to you. Feel free to add your own questions and answer with as much detail as possible.

3. Add your key findings to the holding page (page 33) to use them for brainstorming tactics and actions to include in your action plan.

---

# Visioning: Current Situation

## Part 1 of 3

What does the near future look and feel like for you? Write it and/or draw it using the blank exercise sheet on the next page. Each box contains example questions for you to answer. These are just a few to get you started. You should write as much as you can think of and be as descriptive as possible.

**What does your career or other situation currently look and feel like?**

Are you working in the field?

Do you have an office or a cubicle?

Do you have a regular schedule?

Do you have a say in decisions made?

Will you be able to travel or attend training?

• • • • • • • • • • • • • • • • • • • • • • • • • • • • • • • • • • • •

**What does your personal life look and feel like?**

Do you have weekends off?

Do you have time for hobbies, kids, chores, and other interests?

Do you have time and money to travel?

# Visioning: Current Situation

## Part 1 of 3

What does the near future look and feel like for you? Write it and/or draw it.

**What does your career or other situation currently look and feel like?**

· · · · · · · · · · · · · · · · · · · · · · · · · · · · · · · · · · · · · · · · · · · · · ·

**What does your personal life look and feel like?**

# Visioning: 5 Years Later

## Part 2 of 3

Now that you have a clearer vision of what your near future might look like, imagine that five years have passed. First, picture your career and how it has changed. Be as specific as possible. Then your home life: have you moved, married, bought a new car, etc.? Be sure to take time to envision your future so that you never stop working towards it!

**What do you envision your career will look like in 5 years?**

Will you still be in the field?

Will you have an office or a cubicle?

Will you have a regular schedule?

Do you have a say in decisions?

Will you be able to travel or attend training?

Are you protecting the environment?

Are you in a leadership position?

• • • • • • • • • • • • • • • • • • • • • • • • • • • • • • • • • • • • •

**What does your personal life look and feel like?**

Do you have weekends off?

Do you have time to spend on your hobbies, kids, chores, and other interests?

Do you travel for fun?

# Visioning: 5 Years Later

## Part 2 of 3

**What do you envision your career will look like in 5 years?**

· · · · · · · · · · · · · · · · · · · · · · · · · · · · · · · · · · · · · · ·

**What does your personal life look and feel like?**

# Visioning: Retirement

## Part 3 of 3

If you are a student or just starting your career, it may seem irrelevant to consider this far in the future for your current job search. The reality is: having a clear picture of how you want to exit your career can be one of the most powerful ways to start your career. It will set the stage for the industry you want to get into, the salary level, dictate your level of education, and you who network with. Additionally, as your career progresses, you can use it to assess your level of success towards them throughout your career.

**What do you envision your career will look like when you have the option to retire?**

Will you still be in the field?

What have you accomplished over the years?

What are you known for?

Are you still working full-time or part-time?

Have you saved enough money to reach your retirement goal(s)?

· · · · · · · · · · · · · · · · · · · · · · · · · · · · · · · · · · · · · · ·

**What does your at-home time look and feel like?**

Are you still enjoying work or ready to be retired?

Do you live where you want to live?

Do you have any hobbies?

# Visioning: Retirement

### Part 3 of 3

**What do you envision your career will look like when you have the option to retire?**

- - - - - - - - - - - - - - - - - - - - - - - - - - - - - - - - - - - - - - - - - - - - -

**What does your personal life look and feel like?**

# Skills Gap Analysis

"Always do your best. What you plant now will harvest later."
- Og Mandino

Many job seekers choose to let skills gaps haunt their job search. They will dismiss a missing skill requirement as a fault in the college system or that they didn't get the right degree. Meanwhile, they bullishly apply for positions they are not meeting the basic requirements for, hoping to find an employer who will overlook the gap. We call this the "crossing-your-fingers-approach" and do not recommend it.

We recommend taking a systematic approach to filling the gaps in your resume, so that you will simultaneously be working on Key 1 (Knowledge, Experience, and Interest) and Key 3 (Setting Yourself Apart).

At the end of this exercise, you will have a list of gaps between your current experience and the requirements of the types of jobs that you want. You'll use this list to add to the ideas on the holding page (page 33) to use in brainstorming tactics for your action plan.

## Instructions for the exercise

1. Perform a search for job titles that interest you. Choose at least 3-5 job descriptions to analyze.

2. Scan the job descriptions and look for keywords, repeated phrases, required skills, and recommended skills.

3. Identify the skills where you lack expertise, experience, or confidence. Write those in the left side column.

4. On the right side, write down how you might be able to overcome those weaknesses or fill the gaps. You may have to do some research to complete this.

5. Add your key findings to the holding page (page 33) to use them for brainstorming tactics and actions to include in your action plan.

# Skills Gap Analysis

| Job Title & Keywords | Missing Qualifications | Actions you could take |
| --- | --- | --- |
| | | |

# Key 2:
# Navigating the Industry

## Ensure that you are taking advantage of all your options

*"If you do not know how to ask the right question, you discover nothing." -W. Edwards Deming*

Navigating the Industry is Key 2 in landing your dream career and it means connecting the dots between you and your Knowledge, Experience, and Interest (Key 1) with the possible right fit industries and eliminating industries and fields you aren't interested in. For example, if you want to buy a house, you don't show up to the realtor's office and say "I'll take anything". You have a list of non-negotiables to help narrow down the search. The same concept applies here.

You need to know what your options are to find the best career fit. We are fortunate to live in the era of Google, YouTube, and LinkedIn, where you can conduct all the research you need from the comfort of your home or coffee shop.

Despite some jobs sectors closing, overall the environmental field is actually growing. In the past, businesses made environmental improvements because they were forced to by law. Today, companies are choosing to be environmentally responsible and consumers are starting to demand it, which is creating a positive feedback loop, moving us in the right direction.

This means that doors are opening in many industries. Your goal for this section is to explore what those options are, and make sure you aren't missing out on something that might be a great career for you. We suggest you first go broad in your brainstorming, and then narrow down using your non-negotiables.

Aside from research to understand the industry landscape, you can try internships, volunteering, watching YouTube videos about career paths, contacting people with job titles that you are interested in and asking them a few questions, following businesses on LinkedIn, and more.

**Exercises in this section: Industry Research & Job Research**

# Industry Research

*"Learning what you don't want is how you know what you do want." - Robin Wright*

I often hear defeatist comments like "...but there are no jobs" and "...but I am not qualified for anything".

Despite how it might *feel*, the number of environmental jobs is increasing. According to the Bureau of Labor Statistics, there were more than 90,900 environmental science and studies jobs in the US in 2019 (up from 80,000 in 2017). The number and types of jobs available may vary by location but knowing all of your options allows you to broaden your search criteria and increase the number of opportunities available to you.

For instance, a person applying to "Environmental Scientist" positions only, is missing out on opportunities that someone looking for "Field Scientist", "Sustainability Coordinator", "Hydrologist", and "Biologist" is finding.

Some ways you can research an industry:

· Start with a broad search like "What are the environmental industries?" Then, exclude industries that aren't of interest to you
· Search market reports: look for trends, salaries, and recent developments
· Follow businesses on LinkedIn
· Look up professional organizations
· Follow blogs

### ···················· Instructions for the exercise ····················

1. Before you apply for another job, put some time aside to really explore your job options.

2. First go broad, doing a quick scan and writing down all the possibilities you discover

3. Choose your favorites to research further

4. Narrow it down to your top 1 - 3 to add your key findings to the holding page (page 33)

to use them for brainstorming tactics and actions to include in your action plan.

# Industry Research

Use this space to track your discoveries including new industries identified, professional certifications and memberships worth further investigation, salary notes, trend notes, LinkedIn Groups to follow, businesses to follow, etc.

# Job Research

## "Where there is a will, there is a way." - Proverb

The next research assignment focuses on the types of jobs available to you. You may wish to revisit the gap analysis exercise in the Key 1 section if you discover new positions/job descriptions that interest you.

The goal for this exercise is to ensure that you are maximizing your application and networking efforts by learning about all the potential jobs you might be interested in across the industries that you just learned about in the first exercise.

Some of the ways that you can research available job types:

· Internet search
· LinkedIn search
· Connect and converse with appropriate people on LinkedIn
· Look up companies that interest you and see what jobs they have open
· Connect with people on LinkedIn who have positions that interest you and ask them for a few minutes of their time
· Search job positions on YouTube
· Join environmental and green job groups
· Ask people at networking events about what they do

### •••••••••••••••••••••••••••• Instructions for the exercise ••••••••••••••••••••••••••••

1. Set aside time for this research.

2. Do a quick scan, writing down all the possibilities using some of the suggestions above.

3. Choose your favorites to research further.

4. Choose your top 1 - 3 to add your key findings to the holding page (page 33) to use them for brainstorming tactics and actions to include in your action plan.

*If you have interest in multiple areas, use a plan a/b and even c approach. Keeping a non-negotiable list and search key words for each one. Keep checking to see if each plan aligns with your long term vision.*

# Job Awareness Research

Keep track of your discoveries here including new job titles of interest, job boards to follow, LinkedIn connections who might answer some questions for you, events or groups to network with, etc.

# But What If I Still Don't Know What I Want To Do Or What To Search For?

Not to worry! Contrary to what we see on social media all day long, it is perfectly normal not to know what your calling is. The reality is that most people do not find their calling until after they've experienced something that resonates with them. Sometimes that thing is exposure to a specific need, an event that touches them personally, or a life-changing experience.

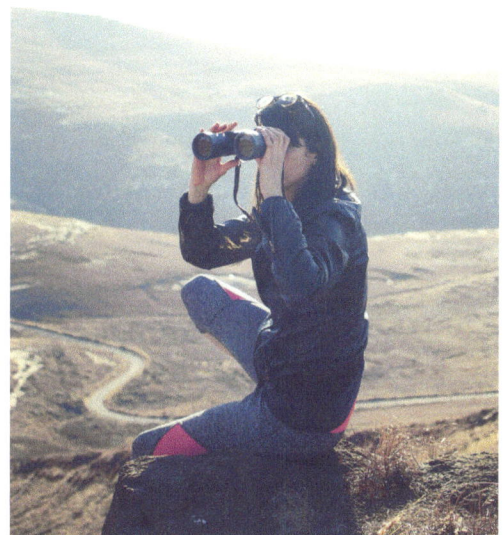

Asking other people what they think you'll be good at or taking a personality test will lead you to clues but you have to continue to reflect on what interests you. I see all the time, people trying to fight against something they gravitate towards because someone else thinks it isn't the right thing for them. Listen to your own compass, then keep moving forward and follow the three E's:

**Evaluate your interest**

**Explore**

**Eliminate**

Keep journaling and revisiting Keys 1 and 2, and the way will make itself clear over time as long as you keep moving forward.

# Key 3:
# Setting Yourself Apart

## Do not give them any reason to pass you by!

*"If you want to stand out in a crowd, give them a reason not to forget you." - Richard Branson*

The final Key in the 3 Keys Model is Setting Yourself Apart. No matter how talented you are, how smart you are, or how outstanding of an employee you would be, if no one remembers you or can't find a reason to put your resume on top of the stack, you won't get the job.

None of the 3 Keys stands alone and Key 3 should be something you think about in all aspects of your job search. Just by having a Job Search Action Plan, you are setting yourself apart!

The exercises in this section focus on you and are broad enough to encompass all that you've already worked in the previous exercises in this workbook. The first exercise here will help you to pull out all that you already know and put it into words. The second exercise is the final exercise in this workbook and is meant to be the catch-all. The personal SWOT analysis will not only reveal some of the gaps you've already identified but also some that you might have missed.

As you put activities and tactics into your Job Search Action Plan, keep in mind how you can continually set yourself apart. Sometimes it is as easy as being the one person out of 10 who follows up or the one person who shows up on time. Other times, you need to be creative and show that you've put real thought into it.

One way that you can do this is to always put the company or other person first. Meaning, always ask: "How can I help them? What can I bring to the table? "

**Exercises in this section: Developing Your Intro Pitch & Your Personal SWOT Analysis**

# Building Your Intro/Elevator Pitch

*"Storytelling is the essential human activity. The harder the situation, the more essential it is." - Tim O'Brien*

Not knowing what to say when you first meet someone is nerve wracking, feeds your negative inner dialogue, and potentially hurts your chances of getting a job you really want!

The purpose of this exercise is to give you a logical format to organize your elevator pitch.

The elevator pitch is your response to questions like, "Nice to meet you, what do you want to do?", "What kind of job are you looking for?" and "What are you in school for?"

Often, when career seekers don't have a good response to these initial questions, they will be very hesitant to start a conversation or to try networking again. Making the right connection in networking is the fastest, best route to your dream career so this is not something you should skip over simply because it is hard to do.

You won't get this right the first time or even the hundredth time. Get a rough draft and then get out there and get networking so you can practice it and continue to improve it. Make a mental note of what works well and feels natural when you are making a new connection. Keep learning and practicing!

························· Instructions for the exercise ·························

1. Read the questions on the worksheet.

2. Write 1 - 2 sentences in response to each.

3. Pull all three of those responses together into a cohesive 1-minute pitch.

4. Practice, Practice, Practice. You're not going for perfection: go for comfort with answering the question "What do you do?" on the spot.

# Intro Pitch Part 1

Use this sheet to build the parts of your 30-second to 1-minute introduction (150-200 words total).

1. List 3 - 5 skills, education, accomplishments, and past/current work experience.

2. State why you are looking for a job in the environmental field. *For example: just graduated, passion for wetlands, want to work out doors, etc.*

3. What type of job are you looking for? (List 3 - 5 Aspects)

4. Put it all together on the next page.

# Intro Pitch Part 2

### Do not give them any reason to pass you by!

Use this sheet to put your 30-second to 1-minute intro pitch together. Practice it and make changes often to see how each version fits. When you find something comfortable that people seem to get, stick with it.

Still having trouble? Do an Internet search or contact a trusted friend, colleague or coach to help you and add the activity to your Job Search Action Plan.

# Personal SWOT Analysis

*"Growth is the great separator between those who succeed and those who do not. When I see a person beginning to separate themselves from the pack, it's almost always due to personal growth." - John C. Maxwell*

SWOT Stands for Strengths, Weaknesses, Opportunities, and Threats. It is a simple yet effective tool that you can use to target tactics and actions to go into your Job Search Action Plan.

The instructions here for a SWOT 2.0 analysis are a little different than you might see elsewhere for a traditional SWOT analysis. First, you'll often see this used as a business tool, but we think it is an exceptional personal evaluation tool. The 2.0 version has each of the 4 categories listed on the left with boxes to the right for you to brainstorm and write down ways you can counter or leverage your SWOT responses.

The SWOT 2.0 is a good way to gauge your progress as well. You may want to update it after you've gotten your first job to see what has changed due to your new circumstances. For instance, a new weakness in a specific skill might emerge or you may find a strength can also be a weakness.

Unlike the previous exercises, the SWOT will give you an opportunity to recognize your strengths and create actions and tactics to take advantage of them. Don't be afraid to write down your perceived and actual weaknesses. These are opportunities for improvement. Align them with gaps you identified in the Skills Gap Analysis exercise.

## ·········· Instructions for the exercise ··········

1. An example sheet and a blank worksheet have been included for this exercise.

2. Find a quiet place and plan to spend at least 10 minutes to answer each question.

3. Try to answer each question as honestly and completely as possible.

4. Add your key findings to the holding page (page 33) to use them for brainstorming tactics and actions to include in your action plan.

# SWOT Analysis Examples

## Do not give them any reason to pass you by!

| Strengths | How can you leverage these? |
|---|---|
| Public speaking | Get some speaking opportunities<br>Attend Toastmasters to get even better |
| People person | Volunteer where I can meet more people<br>Do a better job with networking |

| Weaknesses | How can you overcome these? |
|---|---|
| Afraid of public speaking | Start Toastmasters to practice<br>Attend some networking to get practice introducing myself |
| Not enough GIS skill for jobs I'm looking for | Take more classes<br>Go for certification<br>Start a personal project with ArcGIS online |

| Opportunities | How can you take advantage of these? |
|---|---|
| Upcoming volunteer event | Connect with coordinators in advance<br>Be sure to meet others during the event<br>Follow up after the event |
| Social media | Use my profile to promote my interests and meet people |

| Threats/Negative Internal Dialogue | How can you avoid or eliminate these? |
|---|---|
| "I'm not qualified for anything" | Make sure I'm searching for right jobs<br>Start getting experience through volunteering or courses |
| "What if I don't get a job?" | Work on networking, keep an eye on in-demand skills |

# Personal SWOT Analysis

Do not give them any reason to pass you by!

| Strengths | How can you leverage these? |
| --- | --- |
| | |

| Weaknesses | How can you overcome these? |
| --- | --- |
| | |

| Opportunities | How can you take advantage of these? |
| --- | --- |
| | |

| Threats/Negative Internal Dialogue | How can you avoid or eliminate these? |
| --- | --- |
| | |

# Build Your Action Plan

Congratulations on completing all the exercises! By now you have generated a long list of potential actions you could take to address each of the 3 Keys to land your dream job. These actions will dramatically improve your job search and chances of getting into a career that you will find exciting and fulfilling, but it is unlikely that you will be able to do all of them right now. That said, the next step is to prioritize your list and choose the next set of actions you will take.

Your next step will be to sort all of your ideas into those 3 buckets (Page 34) and choose the top 1 - 3 from each bucket to move forward with on your action plan. The Job Search Action Plan worksheet is organized into three goal-setting sections: networking, applying, and learning.

**Networking**

**Applying**

**Learning**

If for some reason you have struggled to come up with ideas, the next page (page 32) has some suggestions.

Networking is the number one way to accelerate the search system and go straight to "hired." Do not skip it. If it is an area that is intimidating to you, it is more important for you to make goals here. Networking isn't a temporary skill, it is a skill that will help to elevate all you do in life. You can also read *The Environmental Career Coach's Guide to Networking: 5 Keys to Making Life-Changing Connections.*

The applying goals are meant to be things like the amount of time you'll spend job searching or the number of applications that you'll submit per week. Note that this workbook assumes you already have a resume. If you do not, add it to this bucket.

The learning goals help you to fill the gaps on your resume. Remember, give them no excuses not to hire you!

# Suggested Tactics & Actions For Your Job Search Plan

Ideally, you should populate your custom action plan with the ideas you came up with in the workbook exercises to ensure the plan meets your unique needs. However, it can be hard to come up with your own ideas when you don't know what you don't know, so we're providing this list of suggestions. Be sure to give priority to the ideas you came up with.

## Some suggestions to get you started on your Job Search Action Plan:

☐ Evaluate and update resume

☐ Evaluate and update LinkedIn page

☐ Take a Myers-Briggs or 16 Personalities test

☐ Practice interviews

☐ Find a mentor

☐ Read leadership and other helpful books

☐ Attend training and online workshops

## Get creative! What can you add to these?

☐ Hire a coach

☐ Seek internships

☐ Volunteer

☐ Network via LinkedIn

☐ Join a local group or MeetUp

☐ Read a relevant book

☐ Mock interviews

☐ Join a professional group

☐ Get certified

☐ Request informational interviews

A Job Search Tracking Worksheet has been provided on page 37 to help you keep track of where you have applied, remind you when to follow up, and help you to make informed improvements.

Use this page to jot down all the potential ideas you come up with as a result of going through the exercises in this workbook and conducting research. Once the exercises are completed, sort and prioritize your ideas into the 3 buckets of a balanced job search; Networking, Applying, and Learning on the next page.

Use this page to first sort the ideas from the holding page into these buckets and then priority rank them to determine which 1 - 3 from each bucket will make it onto your Job Search Action Plan (page 35-36). Note that some activities and tactics will fit into two buckets. If you try to include more than 3 in each bucket, you may become overwhelmed, which can lead to procrastination.

Priority rank them by considering which will have the greatest potential impact and which you are most likely to implement. If something is easy and only takes a few minutes, add it to your to-do list, rather than your action plan.

### Networking:

*Examples: Reach out to people who work for (these) companies or join a sustainability meetup.*

### Applying:

*Example: Job titles to search for on LinkedIn Jobs.*

### Learning:

*Examples: Take a GIS course or look into certification. Tip: Refer to Skills Gap and SWOT Weaknesses.*

Start by writing your Job Search Goals then list how you will achieve the goal(s) as tactics. This page contains examples. Use the tactics you identified in the exercises in this workbook to populate your custom Job Search Action Plan.

The columns are intended to help you write actionable tactics. Actionable tactics are specific and detailed. Some refer to these as SMART goals, which we do not cover here. You can learn more about creating SMART goals online.

## Examples

**What is your overall job search goal?** *Get an entry level job in marine science*

| Tactic | How can you measure your progress? | When will you do it? | What barriers exist/resources do you need? |
|---|---|---|---|
| *Update my LinkedIn profile* | *Number of sections updated. I will feel confident sharing it with others.* | *One week* | *Time, research a few examples, checklist* |
| *Take a GIS Course* | *Number of modules completed* | *Start in 10 days, finish in 30 days* | *Research options, advice, time, money* |
| *Volunteer with the local aquarium* | *Track the steps through the process volunteering process. Become an official volunteer.* | *Start application tomorrow* | *References, application, patience, time* |
| *Apply for 3 jobs per week* | *Track each application in a spreadsheet and check at the end of each week* | *I will dedicate time Mondays from 1-4pm* | *Don't know what to apply for - need to research (add tactic), time, job boards to search, tracking spreadsheet from the Environmental Career Coach* |

## What is your overall job search goal?

| Tactic | How can you measure your progress? | When will you do it? | What barriers exist/resources do you need? |
| --- | --- | --- | --- |
| | | | |

This bonus worksheet is to help you monitor your application process. Use it as a gauge of what is working. If you're applying to lots of jobs but not getting any return calls, there may be something wrong with your resume. If you're getting interviews, but no offers, there is something wrong with your interview skills or alignment in the jobs you're applying for. You can't properly problem-solve without keeping track. Visit our website for digital and printable templates.

| Employer | Job Title | Where found? | Contact Info | Date Applied | How do you feel about it? | Follow Up Date | Outcome |
|---|---|---|---|---|---|---|---|
|  |  |  |  |  |  |  |  |
|  |  |  |  |  |  |  |  |
|  |  |  |  |  |  |  |  |
|  |  |  |  |  |  |  |  |
|  |  |  |  |  |  |  |  |
|  |  |  |  |  |  |  |  |

**Notes:**

For fillable templates, tools, and other resources
associated with this workbook, visit:
**theenvironmentalcareercoach.com/workbook**

**Notes:**

For fillable templates, tools, and other resources
associated with this workbook, visit:
**theenvironmentalcareercoach.com/workbook**

www.ingramcontent.com/pod-product-compliance
Lightning Source LLC
Chambersburg PA
CBHW052048190326
41521CB00002BA/150